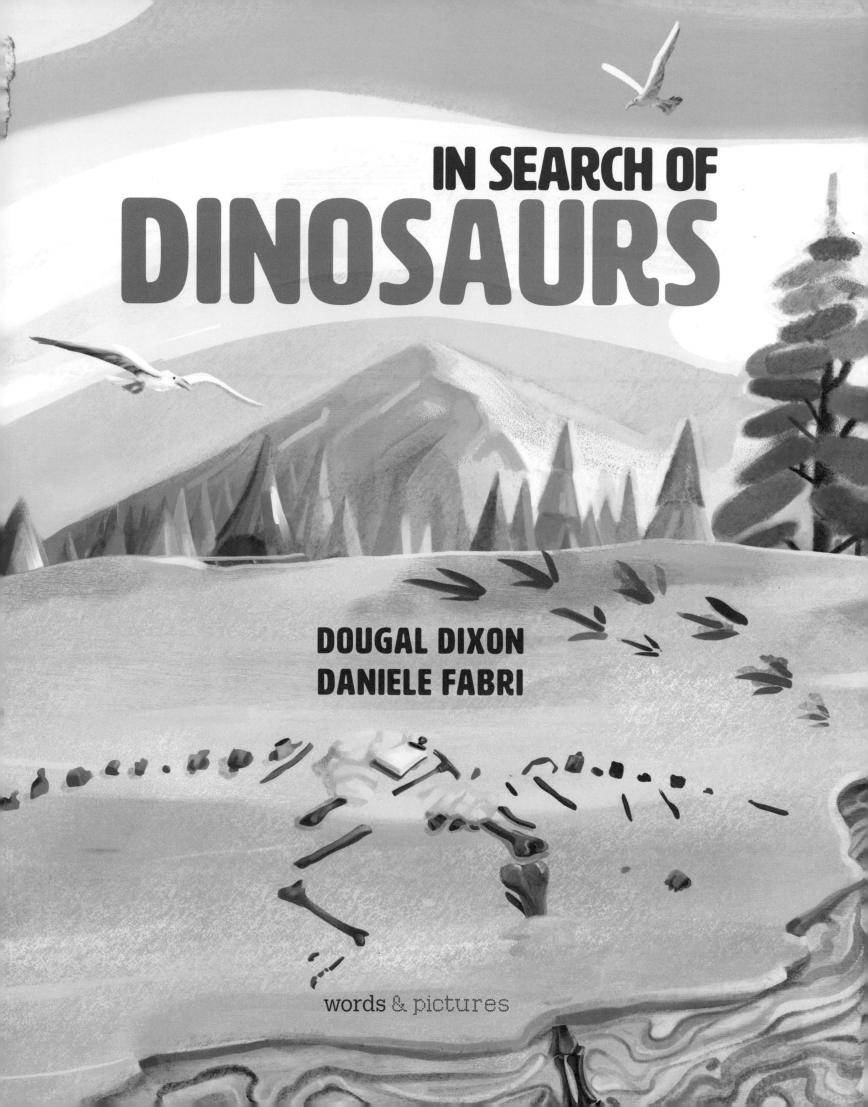

IN SEARCH OF
DINOSAURS

DOUGAL DIXON
DANIELE FABRI

words & pictures

Quarto is the authority on a wide range of topics.

Quarto educates, entertains and enriches the lives of
our readers—enthusiasts and lovers of hands-on living.

www.quartoknows.com

First published in 2019 by words and pictures,
an imprint of The Quarto Group.
The Old Brewery, 6 Blundell Street,
London N7 9BH, United Kingdom.
T (0)20 7700 6700 F (0)20 7700 8066
www.QuartoKnows.com

A catalogue record for this book is available from
the British Library.

ISBN: 978 1 78603 550 9

Manufactured in Shenzhen, China PP032019

9 8 7 6 5 4 3 2 1

Editor: Harriet Stone
Designer: Kevin Knight
Art Director: Susi Martin
Editorial Director: Laura Knowles
Creative Director: Malena Stojic
Publisher: Maxime Boucknooghe

MIX
Paper from
responsible sources
FSC® C001701
www.fsc.org

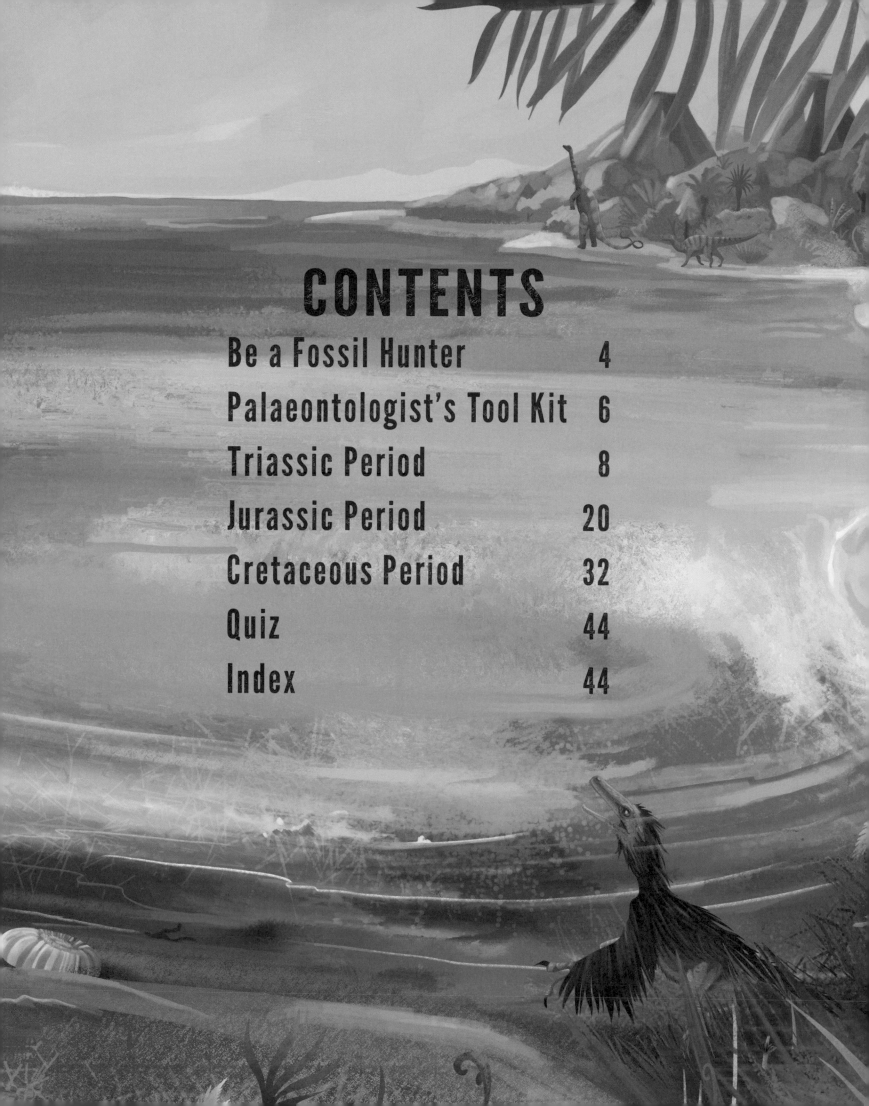

CONTENTS

BE A FOSSIL HUNTER

Dinosaurs were a group of amazing reptiles that lived over 66 million years ago, during the Triassic, Jurassic and Cretaceous periods. They came in all shapes and sizes – but how do we know so much about them?

Their secrets lie in fossils found in rocks all over the world. Fossils form when a creature's body is covered by mud or sand. Over time the mud and sand are packed down and the body inside turns to stone. The hard parts of a creature, such as bones, armour or shells are most likely to become fossils.

Scientists who study dinosaur fossils are called palaeontologists. Spot different types of fossil in this book and see if you can match them to a dinosaur species, just like a real palaeontologist!

TIMELINE

Triassic period
242 – 201 MILLION YEARS AGO

Jurassic period
201 – 145 MILLION YEARS AGO

There are three types of page in this book:

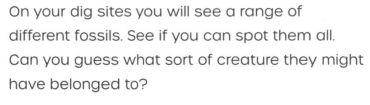

1. FOSSIL DIG SITE

On your dig sites you will see a range of different fossils. See if you can spot them all. Can you guess what sort of creature they might have belonged to?

2. PREHISTORIC SCENE

See the dinosaurs and other prehistoric creatures in their natural habitat. What sort of animals can you spot here?

3. DINOSAUR DIRECTORY

Learn facts about each amazing creature and read clues about what their fossils might look like on your dig site. Can you go back and identify each fossil on the dig?

Cretaceous period
145 – 66 MILLION YEARS AGO

Dinosaur extinction
66 MILLION YEARS AGO

Gather your palaeontologist's tool kit on the next page and you're ready to get fossil hunting!

PALAEONTOLOGIST'S TOOL KIT

Fossils are usually buried deep within a rock. If we want to dig fossils out of the rock to study them, we need special tools. Here are some of the tools that we will use on our dig sites.

Rock can be very hard. We need strong hammers and chisels to remove it from around a fossil.

chisel

hammer

brush

plaster cast

Fossils can be very delicate. We need much finer tools to take them out of the rock. Dentist's tools are helpful for scraping away the rock around small fossils.

scraper

Crumbly rock surrounding a fossil can be cleared away with brushes.

Palaeontologists lay out a grid of string across a fossil site. They use this, along with a tape measure, to note the positions of bones or other fossils and to draw an accurate map of where the fossils are.

string

tape measure

notebook

scissors

small brush

wooden frame

PLASTER POWDER

A very delicate fossil can be hardened by pouring special chemicals onto it. Often the palaeontologist will cover the whole fossil in plaster to protect it, and then take it to a laboratory or a museum where it can be removed and studied more easily.

plaster

bandages

bucket

Now turn the page and see these tools in action on your very first dinosaur dig!

7

This cliffside dig site contains many fossils from the Triassic period. Have a look at the kinds of fossils here, then use the following pages to work out which prehistoric creatures they belonged to.

THE TRIASSIC PERIOD

242–201 MILLION YEARS AGO

Dinosaurs first appeared in the Triassic period. At that time most of the land on Earth was joined together. The earliest dinosaurs lived in small areas within a desert, called an oasis, that had ponds and some plants. Today, dinosaur fossils are found in desert sandstones, such as the rock in our dig site.

Plateosaurus LENGTH: 8 METRES

Long-necked *Plateosaurus* was one of the first big plant-eating dinosaurs. Fossils of *Plateosaurus* are found all over the world.

On our dig site we can see that the *Plateosaurus* got stuck in a swamp – its foot has been fossilized in quicksand. The rest of the skeleton has been chewed up and scattered about. Other animals must have spotted that it was trapped and eaten it!

Megazostrodon LENGTH: 12 CENTIMETRES

Megazostrodon was one of the first mammals to have ever existed. It was small and furry, like a modern shrew, and ate insects and small lizards.

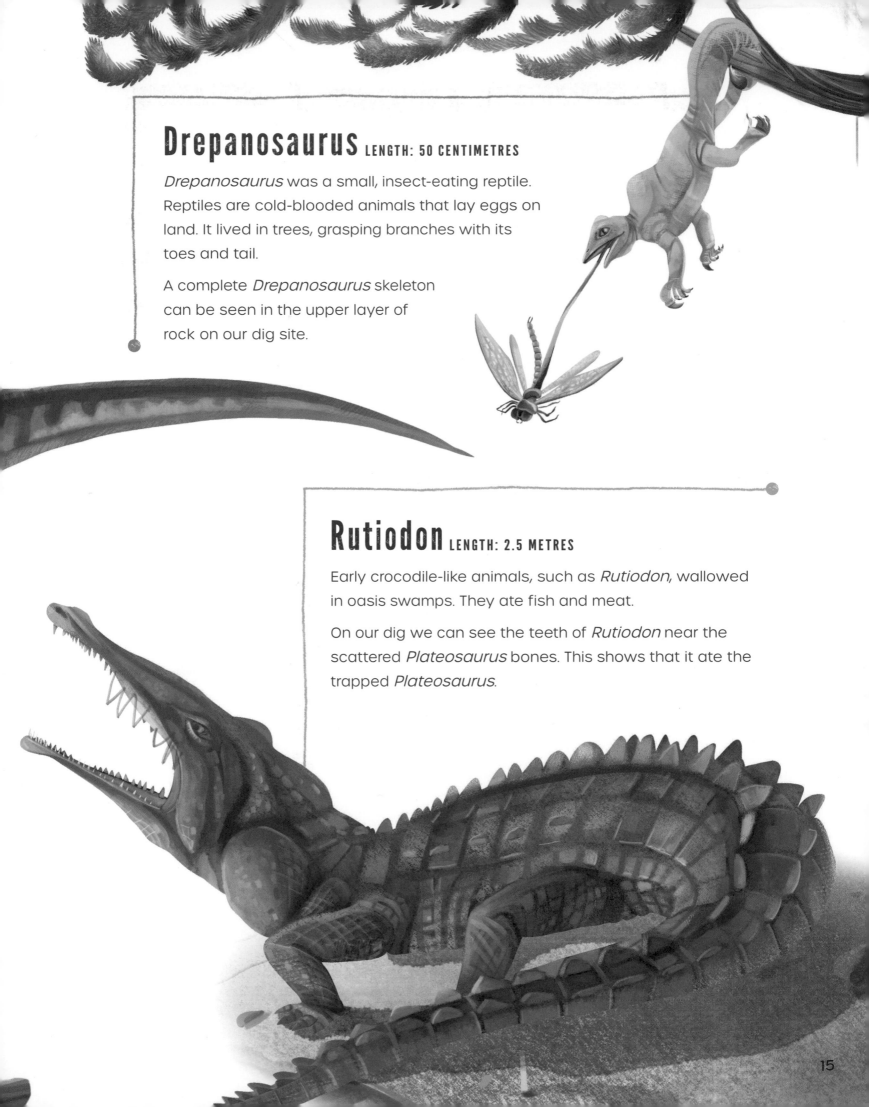

Drepanosaurus LENGTH: 50 CENTIMETRES

Drepanosaurus was a small, insect-eating reptile. Reptiles are cold-blooded animals that lay eggs on land. It lived in trees, grasping branches with its toes and tail.

A complete *Drepanosaurus* skeleton can be seen in the upper layer of rock on our dig site.

Rutiodon LENGTH: 2.5 METRES

Early crocodile-like animals, such as *Rutiodon*, wallowed in oasis swamps. They ate fish and meat.

On our dig we can see the teeth of *Rutiodon* near the scattered *Plateosaurus* bones. This shows that it ate the trapped *Plateosaurus*.

Proganochelys LENGTH: 1 METRE

The first turtles lived during the Triassic period. *Proganochelys* was like a modern turtle, but with a longer neck and tail. It ate only plants and had a beak to crush its food.

There is a fossilized *Proganochelys* shell on our dig site. The shells of this reptile were thick and tough, so its fossils are often undamaged.

Coelophysis LENGTH: 2.5 METRES

Early meat-eating dinosaurs, like *Coelophysis*, were small, quick and hunted in packs.

There are no fossil bones of *Coelophysis* on our dig site, but there are fossilized footprints. *Coelophysis* is likely to have been one of the meat-eaters that ate the trapped *Plateosaurus*.

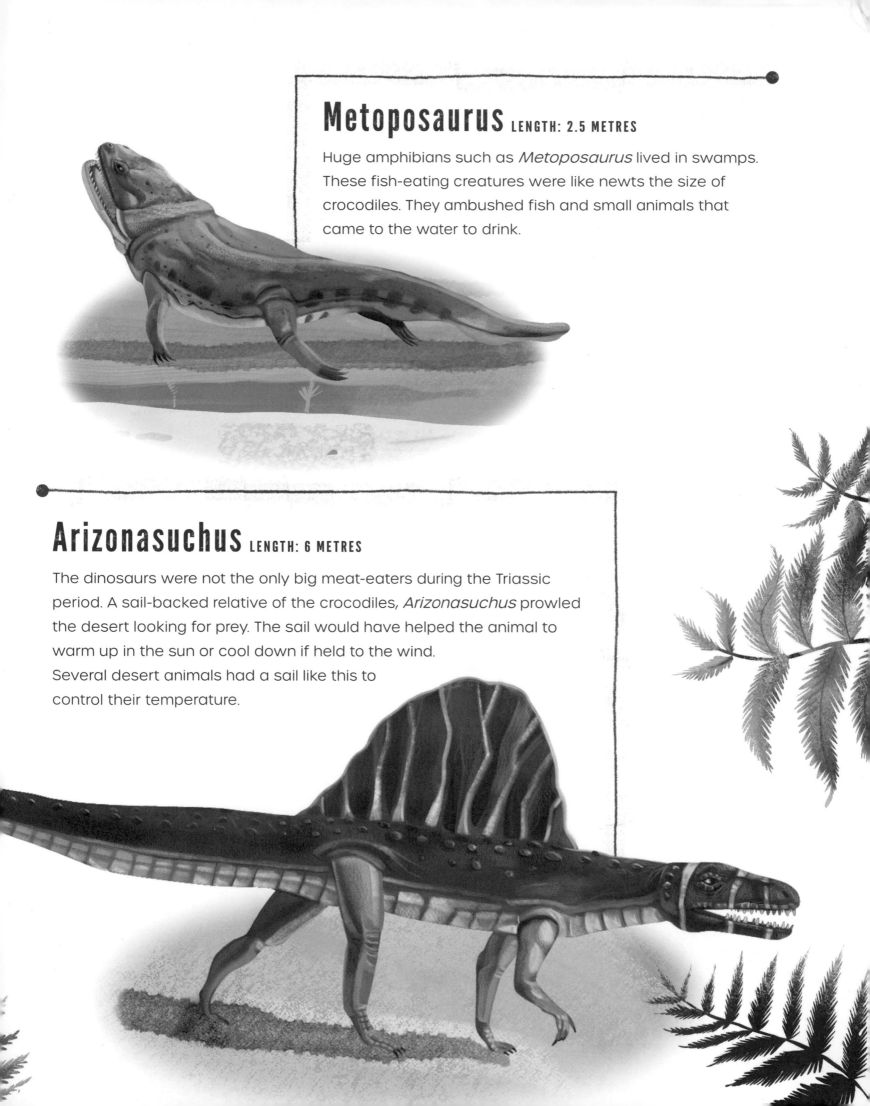

Metoposaurus LENGTH: 2.5 METRES

Huge amphibians such as *Metoposaurus* lived in swamps. These fish-eating creatures were like newts the size of crocodiles. They ambushed fish and small animals that came to the water to drink.

Arizonasuchus LENGTH: 6 METRES

The dinosaurs were not the only big meat-eaters during the Triassic period. A sail-backed relative of the crocodiles, *Arizonasuchus* prowled the desert looking for prey. The sail would have helped the animal to warm up in the sun or cool down if held to the wind. Several desert animals had a sail like this to control their temperature.

Eudimorphodon WINGSPAN: 1 METRE

As well as the first dinosaurs, the first flying reptiles, or pterosaurs, lived during the Triassic period. *Eudimorphodon* was one of the first pterosaurs and, like many others at the time, it ate mostly fish.

A near-complete skeleton on our dig site shows that pterosaurs lived around the oasis. They probably made nests in the surrounding rocks and trees.

Langobardisaurus LENGTH: 50 CENTIMETRES

Little *Langobardisaurus* ran on its hind legs across the desert sands. Wide, strong jaws show that it fed on hard-shelled creatures, such as snails from the land or shrimp from the water.

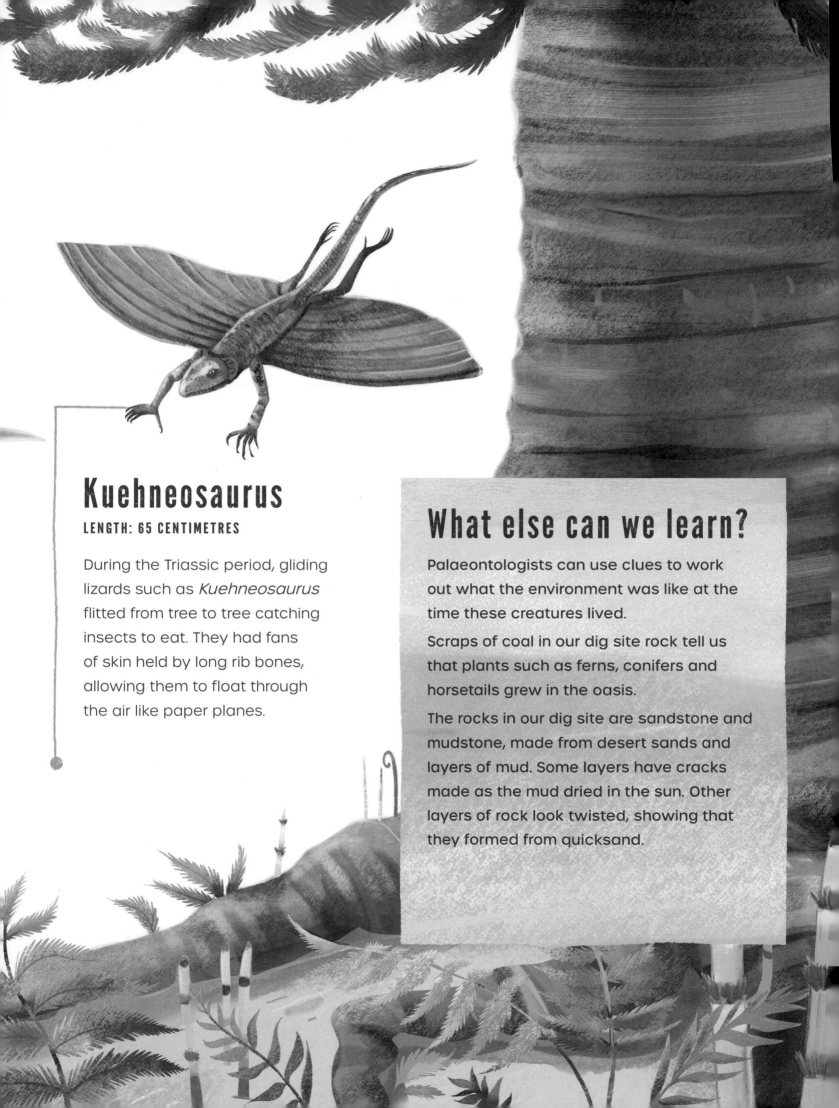

Kuehneosaurus

LENGTH: 65 CENTIMETRES

During the Triassic period, gliding lizards such as *Kuehneosaurus* flitted from tree to tree catching insects to eat. They had fans of skin held by long rib bones, allowing them to float through the air like paper planes.

What else can we learn?

Palaeontologists can use clues to work out what the environment was like at the time these creatures lived.

Scraps of coal in our dig site rock tell us that plants such as ferns, conifers and horsetails grew in the oasis.

The rocks in our dig site are sandstone and mudstone, made from desert sands and layers of mud. Some layers have cracks made as the mud dried in the sun. Other layers of rock look twisted, showing that they formed from quicksand.

JURASSIC DIG SITE

These rock beds were laid down at the bottom of the sea, and then twisted up by the movement of tectonic plates. They contain the remains of ancient sea creatures and animals that died and were washed out to sea during the Jurassic period.

THE JURASSIC PERIOD

201–145 MILLION YEARS AGO

In the Jurassic period the continents started to drift apart. The climate got wetter and there were many shallow seas. Although dinosaurs lived on land, sometimes their bodies were washed out to sea and then buried. Because of this, many fossils are found in rocks that were formed on the sea floor millions of years ago.

Megalosaurus LENGTH: 9 METRES

Megalosaurus was one of the biggest meat-eating dinosaurs of the Jurassic period. Like all meat-eaters, it had a heavy head with long jaws full of sharp teeth. It walked on its hind legs, balanced by a heavy tail.

There is one fossil bone of *Megalosaurus* on our dig site. It is covered with fossilized shellfish and worm holes. This means it must have been lying on the sea bed for a long time after it died.

Juratyrant LENGTH: 3 METRES

There were many smaller meat-eating dinosaurs scampering about during the Jurassic period. They were very active and covered with fine feathers. *Juratyrant* was about the size of an ostrich.

We can see an almost complete *Juratyrant* skeleton on our dig site. Fossils of small dinosaurs are often more complete than bigger ones. Small skeletons get buried in mud more quickly, so there is less chance of them being broken up and eaten by other animals.

Brachiosaurus LENGTH: 25 METRES

Brachiosaurus was a very tall, long-necked plant-eater. Its little head could stretch up to eat the highest leaves in the trees, where no other dinosaur could reach.

There are bones of a complete *Brachiosaurus* leg in our dig site. *Brachiosaurus* was so big that experts once found a huge toe fossil and thought it was a leg bone!

Diplodocus LENGTH: 40 METRES

Long-necked plant-eaters, such as *Diplodocus*, were the biggest dinosaurs that ever lived. *Diplodocus* roamed in herds, feeding on the low-growing plants of the time.

On our dig site we can see hip and leg bones of a *Diplodocus* that have been badly chewed by sea creatures, such as plesiosaurs and ichthyosaurs.

Dacentrurus LENGTH: 10 METRES

During the Jurassic period some plant-eating dinosaurs, such as *Dacentrurus,* had tough plates on their backs.

We can tell that *Dacentrurus* lived on the land near to our dig site, as one of its back plates can be seen inside the stomach area of the *Ichthyosaurus* skeleton. It would have eaten the body of the *Dacentrurus* as it floated out to sea.

Camptosaurus LENGTH: 7 METRES

Smaller plant-eating dinosaurs, such as *Camptosaurus,* prowled the undergrowth of the Jurassic islands. They mostly walked on four legs, but could raise themselves up on two, like a kangaroo, to reach branches and twigs.

Pterodactylus WINGSPAN: 2.6 METRES

The skies of the Jurassic period were full of flying reptiles – the pterosaurs. *Pterodactylus* was one of the most common pterosaurs. It lived close to the sea and fed on fish. It flew using its leathery wings, which were each supported by a long finger bone.

A fossilized wing bone of *Pterodactylus* on our dig site shows that it lived here during this time.

Rhamphorhynchus

WINGSPAN: 1.75 METRES

Rhamphorhynchus was another pterosaur that lived at the same time. We can tell it apart from *Pterodactylus* because of its narrower wings and its long tail.

Stereognathus LENGTH: 10 CENTIMETRES

Stereognathus was part of a group of small Jurassic animals that are the ancestors of modern mammals. It was part reptile and part mammal. *Stereognathus* fed on insects in the undergrowth.

Plesiosaurus LENGTH: 4 METRES

Plesiosaurus is one of many sea reptiles that lived in the Jurassic period. It had a body shape like a turtle and a long, flexible neck.

Fossils of sea reptiles are far more common than those of land dinosaurs, so it is not surprising that there are several bones of *Plesiosaurus* on our dig site.

Ichthyosaurus LENGTH: 3 METRES

The strangest of all the sea reptiles was *Ichthyosaurus*. It looked like a modern-day shark or dolphin, and had a similar environment and feeding habits.

On our dig site there is a complete *Ichthyosaurus* skeleton. It is well-preserved, so must have been completely buried as soon as it died.

Ammonites

Fossilized sea shells are common in rocks like those on our dig site. Many of them come from an animal called an ammonite. Ammonites were like octopuses, but they lived in coiled shells. There were hundreds of different species of ammonite, and each shell was different. They all lived in slightly different eras, so the type of ammonite fossil can tell us the age of the rock it is found in.

What else can we learn?

The rock beds on our dig site are slanted because they have been pushed up by the Earth's moving tectonic plates. If they had not been moved like this, these fossils would still be deep underground where we would never have found them! Scraps of plant material on our dig site show us that the nearby landmass contained ferns, cycads and conifers.

CRETACEOUS DIG SITE

The skeleton of a huge dinosaur lies on the surface of these rocks, which were laid down in riverbeds. But what other animals lived alongside this great creature during the Cretaceous period? Can you spot any of their fossils?

THE CRETACEOUS PERIOD

145–66 MILLION YEARS AGO

The Cretaceous period was the last part of the age of dinosaurs. The land was starting to split into the continents we know today, with different dinosaurs existing on each continent.

Corythosaurus

LENGTH: 10 METRES

Corythosaurus was one of the duckbilled dinosaurs – big plant-eaters with broad beaks. They could walk on two legs but were so heavy that they mostly stayed on all fours.

There is part of a *Corythosaurus* skeleton on our dig site. The skeleton is lying on its side and the rock underneath has an impression of its skin, pressed into the mud millions of years ago.

Dracorex **LENGTH: 3 METRES**

Some two-footed plant-eating dinosaurs had bony domes on their heads for protection. *Dracorex* had a dome that was surrounded by little horns.

There is a fossil bone on our dig site that came from the head dome of a dinosaur such as *Dracorex*, or one of its relatives.

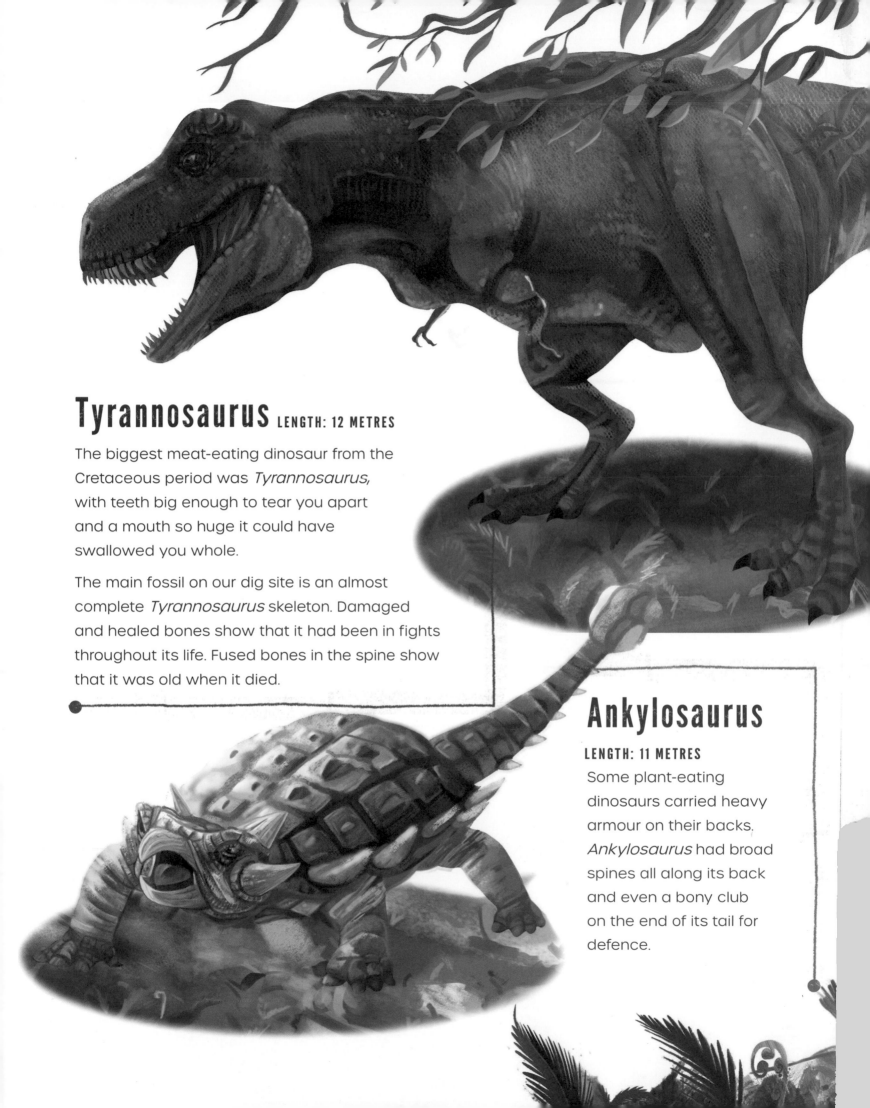

Tyrannosaurus LENGTH: 12 METRES

The biggest meat-eating dinosaur from the Cretaceous period was *Tyrannosaurus*, with teeth big enough to tear you apart and a mouth so huge it could have swallowed you whole.

The main fossil on our dig site is an almost complete *Tyrannosaurus* skeleton. Damaged and healed bones show that it had been in fights throughout its life. Fused bones in the spine show that it was old when it died.

Ankylosaurus

LENGTH: 11 METRES

Some plant-eating dinosaurs carried heavy armour on their backs. *Ankylosaurus* had broad spines all along its back and even a bony club on the end of its tail for defence.

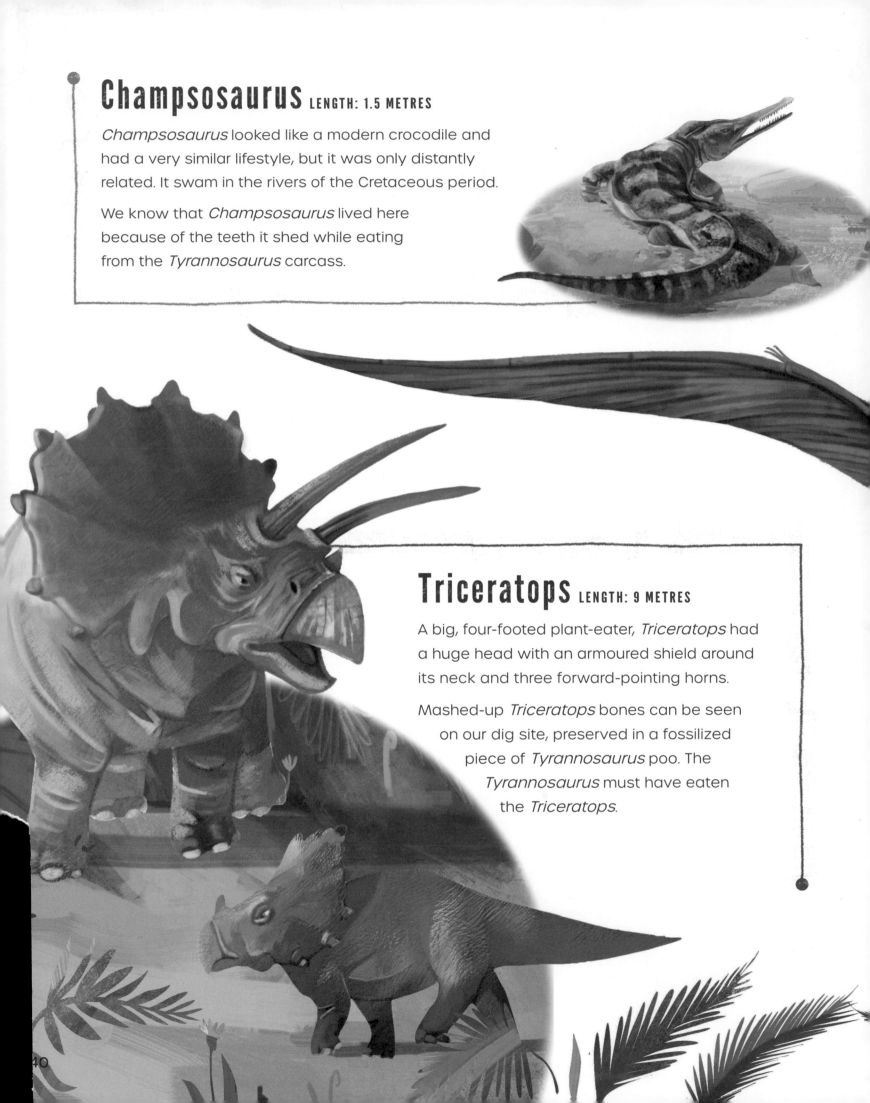

Champsosaurus LENGTH: 1.5 METRES

Champsosaurus looked like a modern crocodile and had a very similar lifestyle, but it was only distantly related. It swam in the rivers of the Cretaceous period.

We know that *Champsosaurus* lived here because of the teeth it shed while eating from the *Tyrannosaurus* carcass.

Triceratops LENGTH: 9 METRES

A big, four-footed plant-eater, *Triceratops* had a huge head with an armoured shield around its neck and three forward-pointing horns.

Mashed-up *Triceratops* bones can be seen on our dig site, preserved in a fossilized piece of *Tyrannosaurus* poo. The *Tyrannosaurus* must have eaten the *Triceratops*.

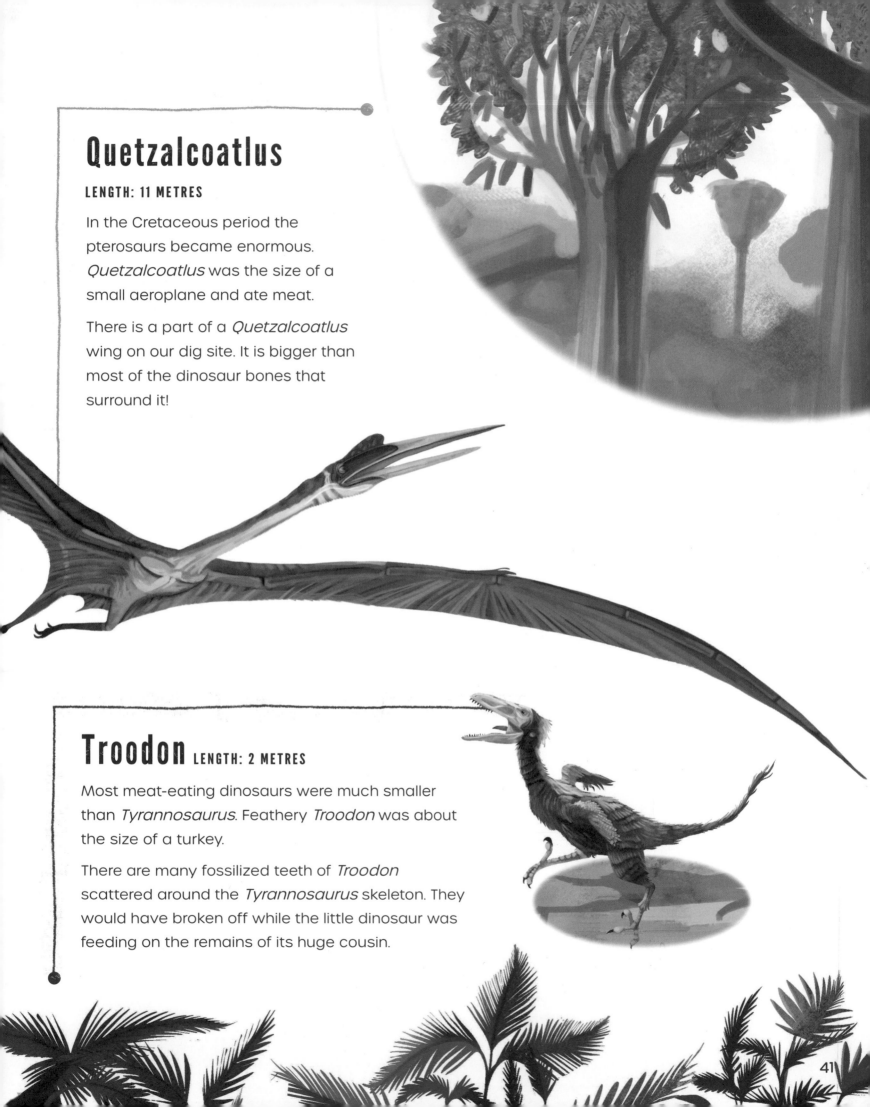

Quetzalcoatlus

LENGTH: 11 METRES

In the Cretaceous period the pterosaurs became enormous. *Quetzalcoatlus* was the size of a small aeroplane and ate meat.

There is a part of a *Quetzalcoatlus* wing on our dig site. It is bigger than most of the dinosaur bones that surround it!

Troodon **LENGTH: 2 METRES**

Most meat-eating dinosaurs were much smaller than *Tyrannosaurus*. Feathery *Troodon* was about the size of a turkey.

There are many fossilized teeth of *Troodon* scattered around the *Tyrannosaurus* skeleton. They would have broken off while the little dinosaur was feeding on the remains of its huge cousin.

Stygiochelys LENGTH: 30 CENTIMETRES

Turtles are an old group of animals. There were plenty of them throughout the age of dinosaurs. *Stygiochelys* lived in the late Cretaceous rivers and ate small invertebrates (animals with no backbones), such as shellfish.

Turtle shells stay undamaged for a long time because they are hard and tough. We can see a well-preserved shell fossil on our dig site.

Cimexomys LENGTH: 20 CENTIMETRES

Bigger than a mouse, but smaller than a rat, *Cimexomys* was one of the little mammals that existed in the trees at the end of the age of dinosaurs. We only know of *Cimexomys* from fossils of its teeth – shaped perfectly for crushing and eating insects.

Small mammals, reptiles and amphibians

Tiny scraps of fossil bone in the rock show that very small animals lived here too – mammals like *Cimexomys*, lizards like *Palaeosaniwa*, aquatic birds like *Potamornis*, frogs like *Beelzebufo* and salamanders like *Scapherpeton*.

Potamornis

LENGTH: 60 CENTIMETRES

Palaeosaniwa

LENGTH: 2 METRES

What else can we learn?

The curved structures in the sandstone show that the original sands and muds were laid down in the bed of a river. The *Tyrannosaurus* skeleton is quite complete, so the body would have been buried quickly before it was damaged too much by scavenging animals. The thin beds of coal show that the surrounding landscape had forests of conifers, magnolias and eucalyptus.

Scapherpeton

LENGTH: 15 CENTIMETRES

Beelzebufo

LENGTH: 20 CENTIMETRES

QUIZ

1 Where is the body of an animal most likely to become a fossil?
A At the top of a mountain
B Under layers of mud and sand
C Up a tree

2 When did dinosaurs die out?
A 100 years ago
B 6,000 years ago
C 66 million years ago

3 What do we use to extract a fossil from very soft and crumbly rock?
A A hammer and chisel
B A brush
C A bulldozer

4 Which of these is a dinosaur?
A Tyrannosaurus
B Pterodactylus
C Stereognathus

5 What do we not usually find as fossils?
A Bones
B Brains
C Shells

6 When did dinosaurs first appear?
A Triassic period
B Jurassic period
C Cretaceous period

Answers: 1 = B, 2 = C, 3 = B, 4 = A, 5 = B, 6 = A.

INDEX